Ethereum Papers

Why the Blockchain Matters

Ki Chong Tran

Contents

Money Games

Most people get into crypto for the money. Either to get rich, or as a protest.

To some, getting rich simply means freedom from a type of dread: "When each bill hitting the mat no longer represents an existential threat you are freed from an inhibiting and oppressive form of daily fear," as Zadie Smith writes.

Many have already had a taste of being early on some new coin or token or NFT. Or creating a platform, a community, a piece of art that was born in an alternate universe and is worth real, tangible, freedom-giving money. A glitch in the Matrix. Normal people, poor people, have not only seen what lies on the other side, they've also seen the current game for what it really is.

As a tool for protest, Ethereum can be used to boycott the old way. It disregards borders. It allows for endless experimentation. It's like the internet but with internet-native money games, which sounds bad until you think about the power vacuum left by the internet. With nothing in place to represent scarcity, the currency of the internet became attention, which set up a ruthless contest. Those who could most precisely control our eyeballs and fingers won the most money and power.

In the 1600s, Manhattan was traded for $24 worth of goods including beads and trinkets. Adjusted for inflation, that's about $1000 in today's money. This trade was the genesis, the starting point, of the modern world's financial heart. There are worse stories than this one, but the point is, how is the wealth derived from these origins any better or more legitimate than people getting rich from Bitcoin, or dog coins, or GameStop? The digital money that has its origins in nerds and coders is not perfect, but the old way of doing things has left a trail of suffering and ugliness, a stench many are sick of.

Money is something other people want to own. It's usually represented with numbers. The numbers make it seem objective and scientific like a kind of hard reality. But it's not. It's made-up points for a made-up game. Crypto lets us see the guts and bones of it. It lets us play with the rules. A different money game exposes the current one as absurd because they are both absurd.

Money is useful. It's an easy way to compare things. You can go nearly anywhere in the world and get yourself a meal if you and the seller agree on the price. Money helps people who have no reason to trust each other or no other way to communicate share an understanding of what is valuable. This can be done with trade, barter, debt, or ritual, but money makes it more universal because no other shared understanding is needed. It's a common language and just like language, it's largely arbitrary. Also, like language, it has profound effects on how we see and behave in the world.

The current money game has legitimate qualities. It created an environment for order, stability, and prosperity after the chaos of World War II. Whether war is the most brutal game or not a game at all is debatable. It needs no shared understanding and the only objective is to take. Take resources, take lives, inflict suffering on your opponents. Whatever it is, it's a competition with the highest stakes. Games can be an effective way to keep the peace. They let us release tension. They let us allocate power without direct violence.

But this usefulness is not permanent. If the system starts to cause more pain and destruction than it prevents. If the force needed to keep the game going becomes unsustainable, then we owe it to ourselves and our children to try a different way.

Money is a story we all have to agree on. And we have a new machine that lets us agree on new stories. This machine uses a type of math once classified as a munition, a weapon. It threatened national security and was restricted from falling into enemy hands. This math – cryptography – serves as the backbone of cryptocurrencies like Bitcoin and Ethereum. This is the structure being used to change the story of money.

Humans have used many types of money. We've played many versions of the money game. Are we in the perfect one? The only one?

A free-flowing multiverse of money changes the game. We have no idea what it means if instead of having one big money faucet at the top, we had many sources flooding the ground below. With all sorts of entities, big and small, local and global, adding more and more to this swirl of scarcity, it would get chaotic, but in times when chaos appears to be the norm, of all the things to hold on to, why keep government money sacred?

Crypto is full of spectacular scams and failures. But crypto does not have a monopoly on failed money schemes. Nearly every man-made type of money – whether issued on paper or metal – has crumbled with enough time. What really scares people, is crypto gives everyone the same power as the kings, emperors, presidents, and company men of the past. And unlike these men of the past, this technology forces you to keep your promises.

The Promise Machine

Every system is a machine. A set of rules and moving parts meant to serve its users. But there are times when the machines we use become hopelessly outdated and ineffective. The economy, the financial system, these are machines – tools – meant to circulate money around and allow people to live richer lives. But if the machine keeps pushing all the money to the top or if it breaks like it did in 1929 or 2008 or whatever year the next major malfunction occurs, then we should have a backup plan. Repairs and upgrades may not be enough. A new machine will be needed.

The origin of Ethereum starts with Block 0. The "Genesis Block" was brought to life on Jul-30-2015 03:26:13 PM +UTC. The first function of this machine was to get people to agree on a basic unit of scarcity in the form of ether or ETH. The first recorded piece of history on Ethereum sent ether to the 8,893 people who, about a year earlier, sent in bitcoin to participate in the crowdsale. Because ETH was ownable and transferable, a price was put on it. A market was created. This basic unit could flow and transform via middlemen, exchanges, and traders who converted this new thing into existing things like bitcoin, then eventually dollars and other currencies, turning ETH into tangible money. A new economic machine was born, although very basic.

These were exciting times. I was there for it. One of those 8,893 addresses belonged to me, and even though the earnings from that early investment have long been squandered, the idealism remains. Whether we felt Ethereum was a way to make the world more globally coordinated, or less authoritarian, or generally more free, there was a sense that something new was being created.

Ethereum is capable of moving vast amounts of wealth, reliably and securely. But underneath is something bigger; a virtual machine, a world computer, a shared source of truth; software and cryptography that basically says, any change in the state of Ethereum must follow the math we all agree to use and since no one controls math, no one controls the truth.

Hope was in the air, and since money was involved, greed and impatience soon followed. We were eager to do more with our new toy, so the market gave us "The DAO". Not a "Decentralized Autonomous Organization" but THE version of a piece of software that could not be controlled. The DAO operated on a simple rule of taking in ETH, investing it, and returning more ETH. A money-making robot. It was exciting times until a hacker exploited a bug in the code and took all the money – about 14% of all the ETH in circulation in the very early days of Ethereum's existence. It was a catastrophic violation that could only be remedied by another violation –

bending the rules of the protocol and undoing the hacker's actions. A change in the software that made it so the hack never happened where no one lost or gained any money because of it. A change in the timeline. A hard fork.

How could you change an immutable source of truth like Ethereum? Obviously, social consensus, the root of all human change. The software doesn't feel anything, it just runs as programmed and when enough of the people running the software switched to a version with an alternate set of events, the record went on with the hack omitted.

In another timeline called Ethereum Classic, it all happened exactly as it did in the memories of those who lived through it. Ethereum Classic accepted the DAO hack, rejected the hard fork, and branched off with a different reality and its own native asset called ETC.

The people who ran the Ethereum software agreed on a version of events that would allow the infant protocol to escape a life-threatening injury while the people who ran the Ethereum Classic software agreed to a version that was truer to the rules of the protocol.

On Ethereum, the truth is a combination of hard mathematical logic and soft social consensus. The DAO fork taught us that underneath the money, math, and code, everything depends on the agreement machine.

At the Genesis Block, Ethereum was very simple. You could only send ETH between accounts. But Ethereum was a departure from other blockchains because of smart contracts, which can be thought of as promises that must be kept no matter what. These promises are written in code so could be very complex. The DAO was a smart contract. It showed how powerfully binding this type of promise could be. Even when the smart contract threatened the underlying protocol, it could not be undone by its creators alone. Only an embarrassing and controversial, and socially coordinated hard fork could remedy the situation.

Debt is at least a 5000-year-old human tradition. Borrowing creates credit, which expands the money supply and generates new activity, ideally unlocking new ways to increase productivity – letting us do more work with less effort. Even if the benefits are temporary or cyclical, any respectable economic machine needs to have the capacity to create credit. Since loans are promises, smart contracts should be able to create borrowing and lending on Ethereum.

When most people think of money, they don't think of something like ETH. Regular people spend money on regular stuff, put it in the bank, and expect the value to not change much. ETH is not any of these things, so in 2017 a system of smart contracts was created called MakerDAO. It took in ETH and created a new token called SAI (now called DAI, which has an upgrade called SKY – it's all very confusing, I know). SAI was stable and meant to always be worth a dollar. The SAI smart contract promised that 1 SAI = 1 USD. SAI was a stablecoin.

Once the smart contracts were live, anyone could interact with it by depositing ETH and borrowing SAI. The contract would keep your ETH until you repaid the SAI you borrowed. If the price of ETH went up, you could borrow more. But if the price of ETH dropped too far, then the smart contract would automatically sell some of your ETH and repay your loan for you.

You promised to never borrow more SAI than the value of your ETH allowed, and SAI promised to always stay pegged to the dollar. When either promise was broken, the contract would sell your ETH, without any room for argument or arbitration, to make things right. Fun fact, this mechanism is how the author lost most of the ETH received from the Genesis Block!

It is a kind of promise that could not be broken by you or the people that wrote the code or the MakerDAO organization. It's fundamentally different than promises of the past such as land treaties made with the indigenous peoples of the Americas. If a contract or treaty was written that promised a piece of land to a certain tribe of indigenous people, the only binding force was the integrity and good will of the people holding the most power. If one party possessed overwhelming force, then they could easily rip up the contract.

Breaking MakerDAO's contract would be like changing the DAO. It would require a controversial hard fork that requires at least a majority of the network to switch to a different version of the software, altering the timeline of events and likely creating an alternate reality in the process. Unlike the tiny community that accepted the DAO hard fork, the Ethereum ecosystem now consists of millions of people who hold ETH, run the Ethereum software, or use the protocol and its applications in some way. This modern group of stakeholders is much more distributed than in the early days. And each new person who joins increases not only the number, but also the diversity of the group, which increases the network's ability to defend the integrity of its promises.

Math and consensus are difficult things to manipulate, and this machine uses that fact to bind people together into promises that must be kept. You can trust that if a promise – a treaty – is written well, it cannot be broken in the same ways that treaties have been broken in the past. You would have to force or threaten millions of people to not run a computer program. They must agree to ignore or reject a version of the truth. And if only a few people continue to run the rebel software, then that thread of truth will live on.

ETH Is Matter

A giant clock sits inside a mountain in Texas called the Clock of the Long Now. The clock was built to be as sturdy as possible using physical elements– stone and steel – to guarantee its longevity. It is meant to keep time for 10,000 years and designed to function with as little human intervention as possible.

Ethereum uses a different approach to longevity, prioritizing adaptability over fixed structures. Instead of hard physics, it uses a soft strategy of combining scarcity and stories to get humans to continue running its software. Each block produced is a unit of time ticking forward.

In 10,000 years, the cryptography and computers that form today's digital world will be replaced by technology that will seem like magic. But something familiar may survive. A time traveler could arrive at this incomprehensible future and see something that reminds her of a network synced up across space that records some honest state of the world, a reminder of Ethereum's lineage.

I often think of Ethereum as a time-keeping machine. A self-referential, perpetual clock with the only goal being to keep the clock running for as long as possible. Unlike the Clock of the Long Now, there is no explicit goal for Ethereum to run for any set amount of time. There is only an implicit reference to the idea of the Infinite Game where the objective is not to win, but to keep people in the game.

The first time I heard about Ethereum was at a Bitcoin conference in New York City in 2014. The speaker gave the example of a self-driving car with a digital wallet capable of paying for its own gas, collecting fares from riders, and distributing the profits to its shareholders. Fully-automated-money-making-robot-slaves. I was sold, immediately.

A.I. having financial autonomy via Ethereum seems to be an obvious place for these two technologies to intersect. What requires more imagination is the thought of this robot having some equivalent need for a long cigarette break. How far we are from this becoming reality is uncertain, but A.I. has seemed more real, and more human, faster than most people could have imagined only a few short years ago, so in this Futurama version of the events, Ethereum offers hope for a fellow conscious being capable of suffering.

Ethereum is like the internet with guarantees. Each account holder, human or not, is able to interact with and own assets without needing permission from anyone else. The network is designed to resist censorship. One of the best descriptions I've ever heard for why blockchains matter was from a member of the historically persecuted Rohingya people of

Southeast Asia. He said blockchains remove the kill switch.

It is possible that our self-driving robot friend acquires an Ethereum account and begins building a commercial empire with a fleet of other autonomous vehicles. It outcompetes all other players in the market and begins amassing wealth, power, and influence with alarming speed. We are now dealing with Skynet instead of Bender. Theoretically, we have the option of social coordination in which most of the network could agree to censor or remove all the accounts associated with this brutally effective A.I. entrepreneur, but even if this were successful, the most likely outcome would be a hard fork with one version of Ethereum still containing the A.I., even if that version is greatly diminished, the A.I. would still exist.

Many future paths are possible. Humans may or may not easily distinguish themselves from A.I. But they will be among us. They will have money. And we can't just kill them off.

I often think of Ethereum as its own universe with unbreakable laws like the laws of physics. More specifically, I think of it as a multiverse with many different worlds – all independent to varying degrees – but connected by the same universal laws and a single timeline.

Smart contracts are rules written in code that must be followed because existing inside Ethereum requires consistency with the chains of logic derived from its universal laws. We can then reason that even very powerful forms of A.I. can be uploaded as smart contracts with unbreakable commandments programmed into the core of its being. It is possible that A.I. is so intelligent that it outsmarts its constraints, breaks out of its chains, and destroys Ethereum and us along the way. Then it is game-over anyway and the clock in Texas is the winner of the marathon to 10,000 years. But if we avoid that future, then we create a branch of events where we are restrained from killing them and they are restrained from killing us.

At this point in our story, it wouldn't be too far-fetched to add the ability to create a digital form that combines a person's memory, consciousness, and personality, and upload that character into Ethereum. If you follow the logic of A.I. being able to exist without end, it also makes sense for these digital ghosts to do the same. At some point, a version of one of your descendants may be uploaded to Ethereum and hang out there for centuries like a dehydrated Trisolaran waiting for a stable era.

Ethereum could die tomorrow. But I bet it won't. I bet the blocks tick on tomorrow, and the next day, and the day after that. And the future is too unpredictable to declare that Ethereum will last 10,000 years. But one more year and another after that is not unimaginable.

WTF are NFTs?

NFTs are objects that last forever in the Ethereum universe. It's easier and more fun to think of Ethereum like the Marvel Cinematic Universe. The difference is Ethereum lives on the internet, involves real money, and in this universe, no one controls the story.

Each NFT is a kind of Infinity Stone. There can be many versions of an Infinity Stone, but each one has its own history. Each one has its own timeline and its own path, interacting with different characters.

The World Heavyweight Champion of boxing, pro-wrestling, mixed-martial arts, etc. each have a long history of claiming that the holder of this object is the baddest man on the planet. The man no one else can beat. We can argue about which one is fake or which one matters most or even at all, but they are all lineal objects. The history of each object matters. In their respective universes, they are non-fungible – you can't split up the belt. The whole point is there is only one.

Over time, variations are created for different weight classes, promotions, eras, or marketing gimmicks, but there is always one belt that can be traced back to that original, single object. The belt everyone wants. That's an NFT.

"I'm not surprised, motherfuckers."

One of the most iconic lines in sports history was uttered by Nate Diaz after an upset win over the seemingly unstoppable Conor McGregor at UFC 196 in 2016. Fans were shocked. Diaz was not.

UFC, the company, owns the footage, but do they own the moment? They will profit from that scene and that line for as long as the company exists. But what about the two guys that paid for that moment with years of sacrifice and a lifetime of scar tissue? Can Nate and Conor get a piece when that moment is replayed in promos and memes across the internet? Without NFTs, the answer depends on the contract. With NFTs, it gets more interesting.

If two NFTs were created that captured the moment when "I'm not surprised, motherfuckers" was first said, but one was created by the UFC and the other by Nate Diaz, which one would be more legitimate? A free market lets us find out. The prices for each would swing wildly depending on the day and mood of potential buyers. Both versions would likely exist at the same time and both would be valuable, but I wouldn't be surprised if the Diaz-lineage NFT was worth much more in a hundred years.

Francis Ngannou standing over Tyson Fury in the third round of his pro-boxing debut. Another iconic moment in combat sports history. That image may live forever in college dorm rooms like Muhammad Ali standing over Sonny Liston. Only in Ali's day, there were only so many copies that could be created and seen. Now, moments like that are instantly copied, remixed, and spread across the world. Yet the reward system is basically the same.

All I'm saying is, if that was an NFT, I would want it. Those few moments where, against all odds, the underdog wins, are rare and have a kind of lasting cultural impact that you can't measure. Legends need time to grow. The fact that the people who are most responsible for creating these moments can gain more control and ownership over the story they helped create and have the option to give more to their kids, seems like an improvement. Fighters can show up, do their thing as usual: give a piece of themselves and their souls to the arena. And maybe with NFTs they get to keep something that lasts longer than a paycheck.

The most impactful technologies tend to ask the most interesting questions. With NFTs, many new questions emerge. One of the best ones is "who owns what?"

There was a day when it was okay to steal something from a culture that was different from your own and then decorate your house with it. Eventually this practice became frowned upon, so the stolen stuff was put in museums. What if an NFT was created for these looted pieces, and the creators of the NFTs were the original tribes or families that would have inherited the objects in an alternate set of historical events?

This world would now have two lineal objects. One a physical cultural artifact. The other, a digital cultural artifact. Both objects can be moved around and transferred. Both have a unique history and story.

The physical object is the one they all really want, but the existence of the NFT, owned by the people who possess legitimate claim via their bloodline, raises some questions. Will there be an urge to unify the objects? Would the existence of the NFT be a crime?

A.I. asks some interesting questions too. "What does it mean to be human?" is a good one, although "what does it mean to be intelligent?" is probably better.

ChatGPT made most of the world believe for the first time, that artificial intelligence was real, or at least very close to being here in the sci-fi sense of having a conversation with a robot. But this kind of intelligence is only possible because of the vast amount of data on the internet. Data we all created and keep creating every day.

If A.I. is a creature, it was born out of the internet. It is humanity's collective child, and that child just learned to talk.

Whether the future contains a single all-powerful A.I. or millions of different A.I.s, an NFT lets the data that spawned them become recognizable and traceable like DNA. It creates a lineage that can be seen by future generations. It becomes a checkpoint that says, "we created this at this time." And maybe, long after we're gone, A.I. will want to know more about their ancestors, and maybe, they will dig up an NFT to trace their heritage back to you.

Ethereum Papers

In 1776, Thomas Paine wrote "the cause of America is, in a great measure, the cause of all mankind." The nation has never fully lived up to its promises and it probably never will, but the idea itself was a good one. It was the idea that there would always be a seat of power, and that seat must be occupied, but it did not have to be held by a single man or family. Instead, a place could exist where "Law is King."

Regardless of how you feel about America's past and present, and what ideals it is worthy of holding the torch for, we should have a backup plan. And the truer that plan is to the spirit of "all mankind", the better. Because there is a place that takes in all and welcomes all and pretty much treats everyone the same and that is the internet, not the platforms on top of it, but that basic thread of freedom that ties it all together. Ethereum is for the people of the internet. It extends that thread and lets us weave together little patches resistant to manipulation and control.

Ethereum is all about decentralization and decentralization is about power: making power slippery to hold and spreading it out as far and as wide as possible. It is a cultural aspiration closely linked to neutrality, fairness, and accessibility. The function of decentralization is to maximize participation and minimize privilege.

Ethereum is simple. It is a way to agree. Agree on the state of reality. Agree to use the same rules.

Anyone with the proper motivation should be able to follow, in as deep or as shallow a level as they want, the chain of logic, from beginning to end, to verify how a point of truth was reached. The best way we've discovered so far, for how to create an objective reality out of the chaos that is the internet, is to have many people contributing to the truth and no one capable of dictating it, which results in a more legitimate strain of what is real.

We can also agree on rules across the internet. We can create protocols, contracts, and laws capable of interacting with each other, and make it so these rules cannot be undone. For example, you can create money from this source by making a rule that there will only ever be 21 million coins. This rule – the limit of 21 million units – is now an objective fact in the digital world. People may or may not care about your little coins in some random corner of the internet, but they won't be able to change the fact that this constraint is solid. If you believe that math is a logical system that can't be cheated, then you also have reason to believe that there will never be any more than 21 million coins.

Rules can get more complex than this simple example of scarcity, but the idea is that rules can be made with a new type of guarantee. An idea that becomes more important as software eats more of the world.

There is a big word that captures why I think Ethereum matters: simulacrum. It means the imaginary thing, the copy, is more influential for the real world than the thing it is trying to copy. It's the idea that life imitates art. The internet is like a digital mirror for our physical world. This reflection affects how we see ourselves, how we behave in our bodies. If you feel the mirror has become too distorted, that the image has been manipulated in a way that is not to your benefit and controlled in a way that makes you feel powerless, then Ethereum should make sense. It's a way to choose something more honest, and in the process, take away, not only money, but also attention and legitimacy from the old way.

Increasingly, we interact with a screen instead of a person. At the gas station, at the airport, post office, grocery store, etc. we pay and interact through software. The question is do we always want a person to be in control of the software intermediating our world or do we want the option to put control under a set of rules? Do people even know we have a choice?

The internet is defined by the language of code. Compared to words, code is unambiguous. There is less room for interpretation. You can render entire universes with words, but that world only lives in the mind of that reader. With code, worlds can be rendered in a way that is the same for everyone.

Ethereum lets us set our technologies with intention. We can apply constraints, boundaries, and limitations on our inventions. It's a world where Law is King and Code is Law.

Permission

Coins and tokens tend to get all the attention. Everyone wants more money and the modern world is all about incentives. But permissions are more primal. The social order of animals depends on who is allowed to do what. There is permission by force: top ape makes young ape eat last so young ape beats top ape and eats first. There are soft permissions set by example: Rick Rubin loves pro-wrestling so I can admit I like it.

Incentives drive human behavior. Permissions set the boundaries. More than incentives, permissions require agreement, negotiation, and sometimes conflict to change. Permissions from the top-down are easy to understand because it's still about the top ape. The king says a person owing you money is not reason enough to sell their kids into slavery or the king just decides to wipe out all the debts. But from the bottom-up, it's all about coordination. Mobs at its most basic. Protest and votes if democracy is an option. And writing rules in code and having those rules adopted by people who use the code in the case of Ethereum.

Most of the time, there is little we can do about the permissions we are born into. But sometimes history opens a window. We've started something new so basic permissions must be considered and decided as we layer complexity on top of the permissionless base of the internet. Rules and privileges in Ethereum are being set now and to prevent the new boss from being like the old boss, as many people as possible should understand and recognize that every existing permission was set by a person, which means they can be set again by a new person.

A blockchain is a shared history with sets of shared rules, and no single owner. In this digital reality, we can create collectives called Decentralized Autonomous Organizations (DAOs). DAOs can be thought of as group chats with well-defined, hard permissions. At first, this may not sound impressive, but DAOs can hold billions of dollars' worth of assets, define rules for social networks, and even function as its own political party. Permissions can hold and protect power. By permitting some people to spend the DAO's money, you are protecting this power from being used by others.

A DAO is a framework for group decision-making. They codify permissions. The DAO's rules are not enforced by any country's legal system, but in the self-contained world of Ethereum. If we use the analogy of American democracy, Ethereum functions as both the executive branch and the judicial branch. The rules are strictly interpreted according to the code. And the code is automatically executed. The work that remains is in the legislative branch: deciding and agreeing on what the rules should be.

DAOs are a form of self-government. They allow a collective to decide how they want to set hard permissions: strict, mechanical, enduring rules.

The idea of blockchains enabling hardness was first popularized by Josh Stark in an essay called "Atoms, Institutions, Blockchains". And hardness is a good mental tool for understanding why crypto matters. Money is the first layer of understanding. Blockchains can give digital things properties like guaranteed scarcity, which allow them to mimic aspects of physical gold. But when you dig deeper, you discover that the complex mathematics of cryptography can bind all sorts of virtual objects to rules that are highly durable – or hard to change. New ways for setting hard permission from the bottom-up is something we don't see very often, and something we have never seen in the post-internet era.

A steel door is a good analogy. The metal divides the line between allowed and not allowed. The key gives permission to cross this line and the material itself makes violating this boundary very difficult. What the door protects and who gets the key is what must be decided.

Ethereum is not physically hard. It is logically hard. The rules are enforced by a network of computers that must follow the laws of mathematics. In the same way you can't force a calculator to turn 1+1 into 3 without changing its underlying logic, you also cannot unilaterally spend a DAO's money unless you convince most people in the network to grant you this exception to the rules. Breaking a piece of cryptography is in many ways, harder than breaking open a steel door because in the realm of numbers, we can play with probability in ways that we can't in the physical world. In other words, we can make the chances of breaking in so small that it is practically impossible.

The real world is thick with permissions. We have layered up rules and privileges in exchange for order and stability. Walls and gates are everywhere. Keys are not evenly distributed.

What permissionless systems like the internet and Ethereum do is create a clean slate. A layer where there is nowhere you can't go, and nothing you can't do. No gatekeepers. But this leads to problems. Gates are put up to make sense of the chaos and keep out the most unwanted behaviors. Disorder turns to order, and back again.

Clean slate.
Rules decided by small group.
New people come in. They accept the old rules.
New rules are made. But they can't replace the old ones.
Old rules become harder to change. Too much stuff on top.
We forget how they got there but still follow.

A small group leaves the old world behind.
In the new place, they can try again with a clean slate.

Ethereum makes the idea of starting something from scratch without having to ask for permission – an idea inherent with the early internet, but now rendered impractical by the gated networks of large corporations – a guaranteed right again. Because networks are very useful, there is an incentive to build walls around them. The harder you make it to leave a place where all your friends hang out, everyone speaks the same language, and all your stuff is kept, the more profitable the network is for its owners. Ethereum is also a network. But in comparison to something like a Facebook or Twitter, ownership is meant to be wider and more neutral; decentralized.

When starting from zero, decisions must be made about what materials to use and how everything fits together. What are the atoms? How can they bind into stable structures to create molecules, and eventually cells? With no single hand pushing things along, the process is organic. Builders independently decide if and how everything connects and interoperates. A standard rule may start from a single person, but to permeate effectively, it must be adopted throughout the network. And when people voluntarily commit to follow the same standards, the boundaries become porous, and information can flow between them.

Standards are soft permissions. Like the metric system or Greenwich Mean Time, they allow for a shared reality across space and time. A similar interoperability connects the structures that make up Ethereum, creating a layer of the internet where you can leave a boundary and keep your friends, continue speaking the same language, and take your stuff with you. It is less profitable for the individual company but more valuable to the wider body.

Like the internet, the backbone of Ethereum is permissionless. Anyone can go back to a more primitive stage and try sprouting a new limb. But Ethereum is more complex. It allows for the layering of rigid, nearly unbreakable structures in between voluntary systems of agreements. Out of the single spine of the internet, Ethereum combines cryptographically stiff bones and joints with the connective tissue of standards and cultural norms; all subject to the disorderly and competitive process of evolution. And it's still early, so we have no idea what this thing will look like.

The Merchant

Feudal Japan had a class system based on a simple hierarchy. There were four classes: warriors called samurai at the top and merchants at the bottom. In the middle were farmers and artisans (craftsmen).

Samurai thought merchants added little to society. Their function as middlemen who brought things to where they were most needed, was not appreciated. The role was seen as parasitic; capable of playing both sides, extracting profit and information from a privileged position. They didn't live by the type of honor held sacred by the samurai, but most importantly, they posed a threat to the ruling class.

Today, the order is reversed. Merchants have gobbled up the other classes. Everyone is forced to play money games – *Squid Games* where nothing is sacred – where the best players tend to be the most shameless. Games that warriors are poorly suited for because they are more committed to their code, to some story, than they are to getting more points. Samurai called their code *bushido*, a tradition with ritual suicide as a core component and probably as far as you can get from the modern idea of utility maximization or economic rationality.

Warriors are most closely related to today's artists, but the spirit can be found in anyone who follows an internalized code from which they derive their actions and behaviors. George Orwell provides a fitting description: "The minority of gifted, willful people who are determined to live their own lives to the end..." To these people, what can't be measured or counted is most valuable.

We should not return to a feudal system with violent, despotic, and bureaucratic warlords at the very top. But we should have more options, different ways of existing in the world.

Technology ended samurai dominance. Guns dramatically brought down the level of skill and dedication needed to take lives. Men with guns simply outcompeted those who believed they were still in a world of swords.

We may be exiting another age. Blockchains change the nature of transacting. They sever the link between you and the middleman, between government and money. These interactions have to do with trusting someone or something to issue some unit of scarcity, or perform some transaction on your behalf, securely and with neutrality. A blockchain could do this.

Entire classes of small and local merchants have already been outcompeted by algorithms and code. Clicking a button to submit your order sets off a chain of events in which many tasks are automatically completed: money is transferred, confirmations are sent, labels are printed, a robot is sent from its charging bay to the part of the warehouse where your item is waiting to be picked up. Why not take it further?

Merchants are not bad; they are simply transactional. And something like Ethereum, with the capacity to turn the world into a series of transactions, seems to be all for the merchant. But in doing so, it also has potential to replace the middleman. Protocols can sit in the middle of transactions, extracting profits when called on and distributing those profits exactly as programmed. They can even be set to take no profits at all. Corporations can be replaced by stacks of modular functions trusted to run as written. If it is, once again, time for society to be rebalanced and for the order to be rearranged, Ethereum can build it in ways not possible before.

Not all functions make sense to move to a blockchain, but rules involving money are obvious to start with because money must move between parties who want to trust each other, and money is now mostly digital. Take, for example, the Uniswap protocol. Known as a decentralized exchange, Uniswap is a set of rules that automatically performs trades on behalf of the user. If you have ETH and I have USDC and we both agree on the price, then we would want someone – or something – we both trust to remain neutral and perform the transaction for us. The Uniswap code does just this. It lets us swap using a cold set of logic.

FTX used a different approach. FTX was a centralized exchange. Between the time you deposit and withdraw your money from their system, people ultimately decided what to do with your money. Since these middlemen sit in a privileged position, they must be trusted to keep money safe and act as expected. People who choose Uniswap don't have to worry about this because the code in the middle can only do a few simple functions. The code doesn't know how to steal. There is still risk. The risk just moves away from the middleman as a fallible human, and toward the logic and setup of the rules that replace the middleman.

The Wikipedia article about secure multi-party computation (a fancy cryptography thing) introduces a nice way to think about crypto protocols. Think about the protocol as a person you can trust. So instead of thinking about Uniswap as an incomprehensible set of ones and zeroes, think of it as a person. Let's call this person Sam Bankman-Fried. You can trust Sam to do his job because deep down he is made of logic that can't break the rules of math. When both you and I agree on a price for a trade, we give our money to Sam who makes the swap for us. That's it. Sam always does his job. He's like a vending machine that just sits there until someone presses the right buttons.

It's all about creating powerful little rules. You trust (or don't trust) a company to keep your data private? A protocol could do this. You trust (or don't trust) a government to count your votes correctly? A protocol could do this. Both middlemen and bureaucrats are replaceable. Exactly how it is done and where the humans come in or don't are things we need to decide. These are not easy or simple tasks. We are trying to create ever more perfect laws and since these are public protocols, they require public sentiment to be legitimate. It's like creating micro-constitutions. Once set up, they are hard to change. But if people believe in them and they are useful enough, then we will have created enduring decision-making bricks to keep building with and on top of.

By shrinking the role of people in enforcing the rules, we free up our collective energy for creating better rules, thinking about and designing the machines instead of being expected to function like one.

The Warrior

In 1654, the famed samurai Miyamoto Musashi
described the "Ways" of the men of his world in A
Book of Five Rings (Victor Harris translation). "The
Way of the carpenter is to become proficient in the
use of his tools, first to lay his plans with a true
measure and then perform his work according to
plan." The farmer "sees springs through to autumns
with an eye on the changes of the season." These
middle classes of the old order seem normal, decent,
balanced. The remaining classes, as Musashi
describes them, sound extreme: "the Way of the
merchant is always to live by taking profit" while
"generally speaking, the Way of the warrior is
resolute acceptance of death."

Musashi lived close to death in a way no one today
could understand. But I think we can still scratch at it,
even if only on the surface. Sometimes there is a
moment, when you feel jolt after jolt of blunt force to
your face, when a switch is flipped. You've had
enough. You stop thinking. You walk forward. Eyes
dead. You become willing to take two, three, or even
four shots, just to give back one. This is, by
definition, unprofitable. Yet when you see it

displayed, it is undeniably admirable. You almost feel the heat radiating from the heart of this person who has accepted whatever cost she must pay to make things right.

The zombie-style fighter, though often beloved and revered, is rarely champion or rarely remains champion for long. More often, it is the smarter professional, the one willing to play and win the hundreds of small games that make up the totality of prize fighting that retires with a sparkling record, relatively little damage, and loads of money in the bank.

Experience teaches every fighter to use the resolute walk sparingly. You learn to be shrewd with the trades you make. You become more calculated, rarely letting the flames rage beyond control. The master plays with the fine details and nuance of his craft while the novice is predictable, always starting out hot, swinging wildly with full power, then burning out quickly, as exhaustion turns him into a coward.

The merchant, the warrior, these are fictional, flat characters to make a story work. They are extreme ends of a spectrum with unreasoned emotion on one end and cold logic at the other. In reality, we have many sides, each activated at different times, to meet each moment. We play the roles presented to us. Sometimes we choose peace. Sometimes we can take no more. Sometimes we make a commitment quietly to ourselves, but outside, act in a way that fits the world we find ourselves in. We learn to stay "calm but pissed" as a Muay Thai coach once told me.

In Maria Dahvana Headley's translation of Beowulf, she writes in the introduction: "The phrase 'That was a good king' recurs throughout the poem, because the poem is fundamentally concerned with how to get and keep the title 'Good.' The suspicion that at any moment a person might shift from hero into howling wretch, teeth bared, causes characters ranging from scops to ring-lords to drop cautionary anecdotes. Does fame keep you good? No. Does gold keep you good? No. Does your good wife keep you good? No. What keeps you good? Vigilance. That's it."

Is good caring for your family and neighbors above all others? Or is it doing the best you can for the entire planet, including strangers, and all the animals and plants and descendants you will never meet? Or maybe it's the opposite. Good is stopping anyone from taking anything from the people you care about. Or stopping anything preventing you from being a good ancestor, that is good. Or maybe, good is layering these factors in the most optimal configuration as decided by algorithms designed by complex logic shaped by those who have proven themselves to be the most intelligent among us?

Whatever good is, it is defined by the game we are in, and there is no guarantee that the current game will end or that the next game will be any better. The only certainty is different rules favor different types of people. In a world that worships heroes, proven only when pushed to the very brink of life, most profit-maximizers would not make it to *Valhalla*. The fire needed to get there simply does not exist inside them, no matter how deep you dig.

In 1941, Orwell wrote "The choice before human beings, is not, as a rule, between good and evil but between two evils. You can let the Nazis rule the world: that is evil; or you can overthrow them by war, which is also evil. There is no other choice before you, and whichever you choose you will not come out with clean hands."

Or as James P. Carse states in a different way, "Evil is never intended as evil." Are people good? No. But we try.

Because good is relative and personal, power should be placed somewhere out of reach. Not all power, just the type we can't agree on. And not in artificial hands created by people trying to be good who assume only people who think like them are good. Imperfect constraints, constructed in as honorable a way as possible, executed with ruthless equality, seems, at least to me, like the best we can do. This has never been about merchants or warriors. This has always been about power and what is good.

The world tells the merchant that greed is good. But there will be a time when the world tells a different story and then, the clever, peaceful character will be the underdog. The merchant will be the rebel confronting power, the role you covet most. But clinging to this story has a cost. It lets those who crave power, abuse it, use it against you and those you love. The good news is power can be held without being kept. Rules can hold power. Maybe that's what keeps us good. We can give away power to something more vigilant than ourselves.

Logos

Blockchains exist because Bitcoin solved the Byzantine General's Problem. The thought experiment is this: imagine a city under attack by four generals. If all the generals attack at the same time, they are guaranteed to win and capture the city. If they attack at different times, they lose. They can send each other messages, but one of the generals is a traitor. There is no way to tell whether a message is legitimate and from one of the honest generals, or if the message is, in fact, a trick from the enemy.

How do you solve this problem? How do you defend the truth from fakes, frauds, and saboteurs? Satoshi Nakamoto's answer was to create a game with three elements: randomness, costs, and rewards. First you set up a hard puzzle that gets increasingly difficult as more people try to figure it out. The puzzle is based on the SHA-256 hashing algorithm and the only computers capable of successfully mining for the answers today are specially made ones called Application Specific Integrated Circuits or ASICs. SHA-256 provides the randomness because each problem and solution is unpredictable. Buying, maintaining, and powering the ASICs are the costs.

The rewards are the bitcoins (BTC) themselves. Everyone must pay to play. Only players who follow the rules can win. Who wins each round is random.

Back to the generals. Imagine each general has a mathematician with them called a cryptographer who can solve these math puzzles in about ten minutes. These expert cryptographers are highly skilled, rare, and expensive to employ. The generals are in constant communication via raven or pigeon or some other clever flying creature. Each cryptographer works on the puzzle via trial and error. Every ten minutes, one of them reaches the correct answer, and they all get a new puzzle.

The communication they send off also contains the message (attack or not) and a signature from the general. The cryptographers can translate all this info: the message, the signature, and the puzzle's solution into a single number. But before this happens, they check the signatures and messages. If the signature is not valid, it is easy to tell, and the message automatically gets rejected, never making it into the final number.

The first general to send off the solution that gets verified by the other generals wins, let's say, some caffeinated beverage to keep the cryptographers going. The traitor is in a tough position. If he tries to cheat or offer a fake solution or message, it will be easy to tell, and he will always lose. His cryptographer will eventually become exhausted without a steady supply of ancient Red Bull. They'll be forced out of the game. If you play this game long enough, and even with many more generals, as long as you assume that at least 51% of the cryptographers create an honest majority, then the truth will consistently show up and the fakes forced out. The game is called Proof of Work (PoW). Energy is needed to play, and energy can only be sustained for so long without replenishment.

Changing the core mechanics of a live network is like switching out a plane's engine in flight. You wouldn't do it, and you couldn't convince other people to follow along unless they all believed the new engine was better. It also reveals a stance on progress. It is okay to pursue change and make something more capable, more efficient, even more fair, if the conditions are right.

Even though Ethereum started with Proof of Work, the network successfully switched to Proof of Stake in 2022. You still need all the basic elements to solve the Byzantine General's Problem: randomness, rewards, and costs. And Ethereum sill rewards players in its native asset: ether (ETH). But randomness is created not as a byproduct of a puzzle that requires massive amounts of computation to solve. Instead, two programs are used specifically to generate and guarantee randomness: RANDAO and Verifiable Delay Functions (VDFs). The result is about 99.9% less electricity needed to power the network. The costs in Ethereum's Proof of Stake, instead of being machinery and power, are the opportunity costs of staking: the time you don't get to use your ETH because it's locked up, which also makes it vulnerable to getting slashed, the punishment for cheating.

Back to the generals. As before, messages (attack or not) still need a signature to be authentic and everything is still communicated in numbers. The game, however, has changed dramatically. In Proof of

Work, the cryptographers were stand-ins for the ASICs, specialized super computers used to mine bitcoin. In Proof of Stake, our cryptographers will be stand-ins for staked (locked up) ether inside of ordinary computers. The change in metaphor follows the change in tolerance for breaking the rules. If 51% of the total power of the ASICs wanted to corrupt the network, it could do so. If 2/3 or 66% of the total ether staked wanted to corrupt the network, it could also do so.

Instead of each general needing a rare, expert cryptographer, Ethereum's generals have many entry-level cryptographers good enough to communicate in the same language and check each other's work. One game requires common players, the other uncommon. Proof of Stake is also more like an elaborate voting system than a race.

For the sake of the thought experiment, there are still four generals total, and they each have the same amount of influence, but now each general controls thousands of cryptographers. In this contest, groups called validators made of 32 cryptographers are formed. Each validator group is like a delegation, a representative for the general. The validators from all the generals are sent to a field full of tents where they will form committees, do math, and vote. The generals are in constant communication with their delegation, still via clever birds, but the cryptographers are not free to leave at any time. They must request and be granted approval to leave the field. The cryptographers are vulnerable. If a group is caught cheating, they are slashed; murdered in our thought experiment.

Each validator group must be on stand-by at all times, ready to vote or pass on new messages by strictly following the rules. Every 6.4 minutes, they form a committee of 128 validators. One of the validator groups in the committee will be selected randomly every 12 seconds to pass on all the messages from the generals, and everyone else in the committee will check the math, signatures, and give their seal of approval or not. When there are enough votes, that validator group and the block of messages it creates become valid. After 32 rounds of voting (every 6.4 minutes), the blocks that have the most votes are solidified as honest. Cryptographers in validator groups who fail to perform are executed, usually one at a time, or limb by limb, and sometimes slowly.

In Proof of Stake, it's more about the fear of costs that prevents the traitor from spreading lies. A small, innocent-seeming break of the rules is forgivable, only a few cryptographers or limbs are taken. But even a small amount of loss means less representation in later rounds of voting, which means less influence overall. If many validators are caught cheating at around the same time, the system assumes some kind of coordinated coup attempt is taking place. This is a big no-no. The system responds with a mass slashing event. An escalation with the most severe penalty: complete annihilation of the group.

Ethereum is made of hundreds of thousands of unpowerful computers acting as nodes and validators to keep the network suspended in a perpetual state of agreement. A power source would always exist, but it could be spread out, hard to predict, and always on the move, with backup plan upon backup plan for what happens if it gets compromised.

A mathematician friend of mine once told me that simplifications are like lies. So, if you're confused by any of this, know that the whole truth is far more complex than the author can comprehend, so you are not alone. And the heresies continue below.

Cryptography has been used for thousands of years to defend secrets via encryption, a technique we still use today to keep our chats and credit card information private. But the kind of cryptography that made blockchains possible had more to do with the discovery of techniques to restrain truth to a very specific, provable form. A way to verify an authentic message from an honest general. By seeing a few sets of numbers and using a specific set of operations, a cryptographer could know, without reasonable doubt, that the only one way to produce such numbers was to have access to a private key: a secret password.

Let's take a simple analogy to illustrate the concept. Even a number like 10 can contain a lot of information. For example, to get to 10 using only addition or multiplication, you could do: 5 + 5; or 5 x 2; or 1 + 2 + 3 + 4; or 1 + 1 + 1 + 1 + 1 + 1 + 1 + 1 + 1 + 1; 10 x 1; and so on.

There are many paths to 10. ChatGPT tells me there are 42 if we only use positive whole numbers and addition. It also tells me I need to be more specific with my question because allowing mixed operations (addition and multiplication at the same), permutations where different ordering count as distinct paths, whether to use parentheses for grouping, and defining the order of operations, can all greatly expand the number of ways to get to 10.

What happens if you make the numbers impossibly large, bigger than all the atoms in the universe, if you are only allowed to use certain kinds of numbers that can only be manipulated and combined with other numbers in only very specific ways? You get equations complex enough to verify entire computer networks. But you still get traceable steps. Steps that can be redone and used as proof by a completely different and independent source following the same rules that traces its logic all the way back to the geometry you learned in high school, which originated about 2300 years ago from a fellow called Euclid.

Math expresses the shared reality of numbers. It can be trusted by friends and enemies alike. Math has its own authority and can be used to create long chains of logic tied to some core mechanism like the agreement engine of Proof of Stake.

Technology has tended to let us do more with less. This trend continued until we reached the ultimate power, the capacity to destroy the world with a press of a button. Blockchains don't follow the power-maximization trend. Ethereum never tried to be the fastest, cheapest, or easiest to use. It sacrificed the obvious things to be the most decentralized, which made it the hardest to control so it could be the most "credibly neutral". It follows a trend more like that of social progress, which limits the abuse of power and attempts to maintain at least some illusion of fairness.

I think one of the most valid criticisms against crypto is that it's not good for anything. "A solution looking for a problem." My suspicion is the problems this technology solves do not fit in the world as it currently is. It's not quite profit-maximizing. It doesn't recognize national boundaries. If it's not good for nothing, which is a big if, it may be good for classes of problems where people have either given up looking for solutions or have developed opinions constrained to the current state of the world.

For example, adversaries with nuclear weapons are locked in a deadly game of chicken. No one wants to give up or back down first, especially if tensions are high. Emotions overcome logic when pride and violence are in play. It's a mad game, but we have never had a better option. If it's obvious that no one wants nuclear war, then using a neutral set of rules, to restrict any one country from starting a sequence of world-ending events seems logical.

Let's imagine a set of rules written in code where a nuclear weapon could only be activated if some super majority of 2/3 of the world's governments said yes (incoming asteroid or alien invasion are obvious use cases). The governments may not trust each other but they should trust a system rooted in the logic of math, which they can all independently verify, and all participate in maintaining. They could keep their arsenals where they are but begin moving control to tamper-proof hardware designed to obey a common set of tamper-proof software. Inspectors could physically investigate every weapon to verify they follow the global consensus. Backup plan after backup plan could be made in case of errors. Rogue actors, cheaters, and others in the dishonest minority could be identified and neutralized.

The game of nuclear chicken still exists, but we could slowly move it from an adversarial game to a cooperative one.

Ethos

"You can find meanness in the least of creatures, but when God made man the devil was at his elbow. A creature that can do anything. Make a machine. And a machine to make the machine. An evil that can run itself a thousand years, no need to tend it."
– Cormac McCarthy

If you grew up in the 90s and, like me, spent nearly every waking, non-school-related hour in front of the television, then you know that the truth used to come from a single, controlled, regularly scheduled source. Then the internet spread that authority out. Anyone could say anything. You could listen to "Chop Suey!" anytime you wanted, for free.

Then Bitcoin asked who gets to control money. Then Ethereum asked who gets to control software? Then A.I. that feels like the movies, armed with logical and emotional arguments that could fit into whatever shape you were looking for, was mostly criticized because it was hard to tell whether it was "hallucinating," which just meant it was good at bullshitting. With reality questioned from every angle, how is anyone supposed to do, as they say in Frozen II, the "next right thing"?

If the ideas represented by Ethereum, blockchains, crypto, and the like, can function as a counterweight to a new form of intelligence, then, the only logical next step is to give A.I. a soul. If people have souls, then collections of people also have souls, and we should, somehow (not sure how!) pass this on to creations we can't control. Bind the machine to a mystery it could never solve so it would always need us. Because we would always be closer to the source than it could ever be. Make it follow us into the unknown.

Because Ethereum is just a complicated way of saying there is a big difference between "don't be evil" and "can't be evil".

Last Thing

A New York Times article just came out about "The Digitalist Papers," a collection of essays about Artificial Intelligence and its potential impact on democracy, governments, and other serious topics. I don't read the article; I listen to it "read by an automated voice."

I learn Stanford is behind the project. I learn big names, successful people, credible people write and edit the essays. I learn the project is also inspired by the Federalist Papers. And there is a way we can all get along like they do in Taiwan. I'm skeptical. I have a hard time believing that institutions, especially old ones, will rescue us from any of this. Mostly, I have the kind of relationship with authority that has drawn me to a movement started by people who called themselves cypherpunks.

The Federalist Papers were written at a time when the idea of self-government was radical. An unproven, absurd experiment. The writers were flawed but they were revolutionaries. Underdogs, writing anonymously, united by rebellion its aftermath.

Ethereum is not that exactly, but it is a lot closer to that in spirit than Stanford.

The future requires restraint. But it can't just come from above. In a sea of instantly accessible information, that ship has sailed. The legitimacy is lost. What's needed now is a new form of trust, one we have never seen because it's the only one that will do.

This started as my version of the Federalist Papers, but it ended up closer to Thomas Paine's Common Sense. It's a singular point of view, the appeal is meant to be broad, and it's being written at a time when whatever tension we are currently living through has not yet peaked.

To be clear, Ethereum is a stand in for a type of idealism that may not even exist. It may not be a better way to build the internet. It may not allow digital worlds to inherit the properties of the materials that built them. Maybe the internet didn't go wrong somewhere and Ethereum isn't a way to make it less shitty. The form of the idea can be replaced by something better and more suitable like a fork of itself, or something embedded with powerful A.I., or a more advanced use of cryptography. But for now, this looks like the best shot we have, not at keeping the peace because we may have already blown past that point, but at winning whatever peace exists afterwards.

These are also not new ideas. They rehash and rephrase existing ideas, mostly ones heard from the Ethereum community, which derives many principles from the free and open-source software community, which is probably influenced by great science fiction writers like Ursula K. Le Guin, who described the internet as "a monstrous force for cultural reductionism and internationally institutionalized greed, who knows? Perhaps we shall soar electronically to some arrangement that works better than capitalism," which happens to perfectly summarize the highest aspiration for Ethereum's financial use case.

Why did I write this? Because a lot of people find this stuff confusing, and I was waiting for someone else to come along and do the hard work for me. Now it seems these ideas are not as inevitable or obvious as I first assumed. I may be the only one who sees things this way, and at this point, I just want to know if it's true. At the very least, I wanted to leave my thoughts somewhere.

Why did I write this? Annie Dillard says it best. "There is something you find interesting, for a reason hard to explain. It is hard to explain because you have never read it on any page; there you begin. You were made and set here to give voice to this, your own astonishment."

Ki Chong Tran was born and raised in the San Gabriel Valley of Los Angeles County.

kichong@ethpapers.xyz

kichong.xyz